Reach for the Stars

"How I became an astronaut" – The story of Stephanie D. Wilson

by Stephen Schutz*

*Written by Stephen Schutz based upon interviews with Stephanie D. Wilson

Starfall® Education
P.O. Box 359, Boulder, CO 80306

Copyright © 2009, 2012 Starfall Education. All rights reserved. Starfall is a registered trademark in the U.S., the European Community and various other countries. Printed in China. ISBN: 978-1-59577-127-8

Stephanie at age 6 on a family vacation, Niagara Falls, NY

Curiosity

I was just like any other child.

I was very curious, liked to travel, and always asked a lot of questions.

Horsehead Nebula

Enthusiasm

In 8th grade, a homework assignment changed my life.

Our teacher asked us to talk to grown-ups with interesting jobs. I spoke with an **astronomer**.

The astronomer was bursting with **enthusiasm** as he talked about the millions of stars in the sky. That's when I became interested in space and decided I wanted to be an **astronaut**.

Graduation in 1992 with Master of Science in Aerospace Engineering from the University of Texas at Austin

Education

I worked hard in school. My parents and teachers encouraged me to have faith in my abilities, and to follow my dreams.

My favorite subjects in school were science, languages, and music.

I also became more interested in space travel.

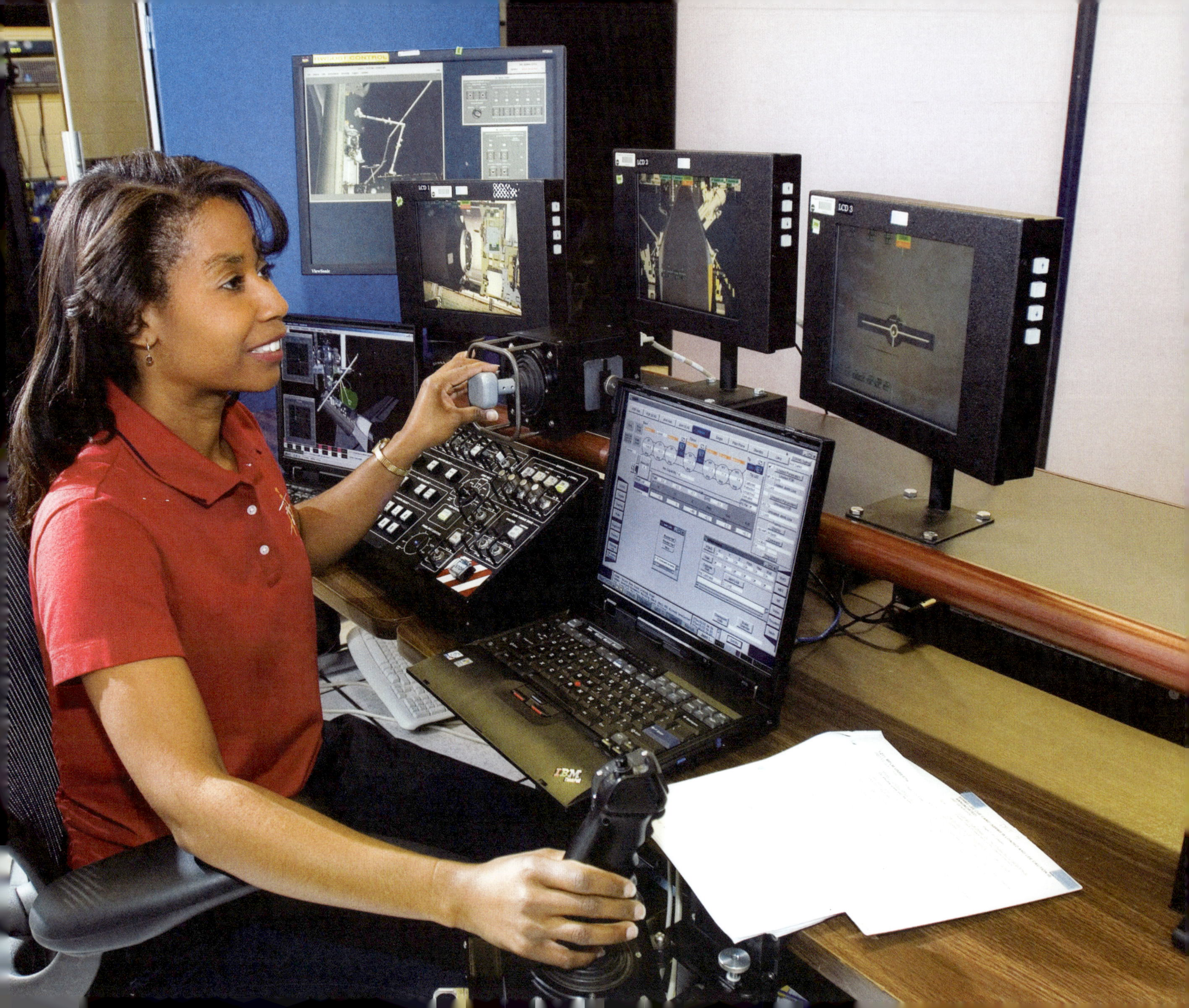
LCD 1
LCD 2
LCD 3
IBM

Practice

After college, I was honored to be chosen to join the astronaut training program at **NASA**.

As I trained, I practiced many new skills over and over again until I could do them well.

Crew of Space Shuttle *Discovery*, Mission STS-120

Excitement

At last I was ready to go into space! On the day of the launch, I was very excited.

I walked out of the building where we lived for a few days prior to the launch, wearing a 35-pound **launch and entry suit**. I had exercised to stay strong and ate healthy foods including fruits and vegetables. That's why I had plenty of **energy**.

The **space shuttle** needed a different kind of energy…

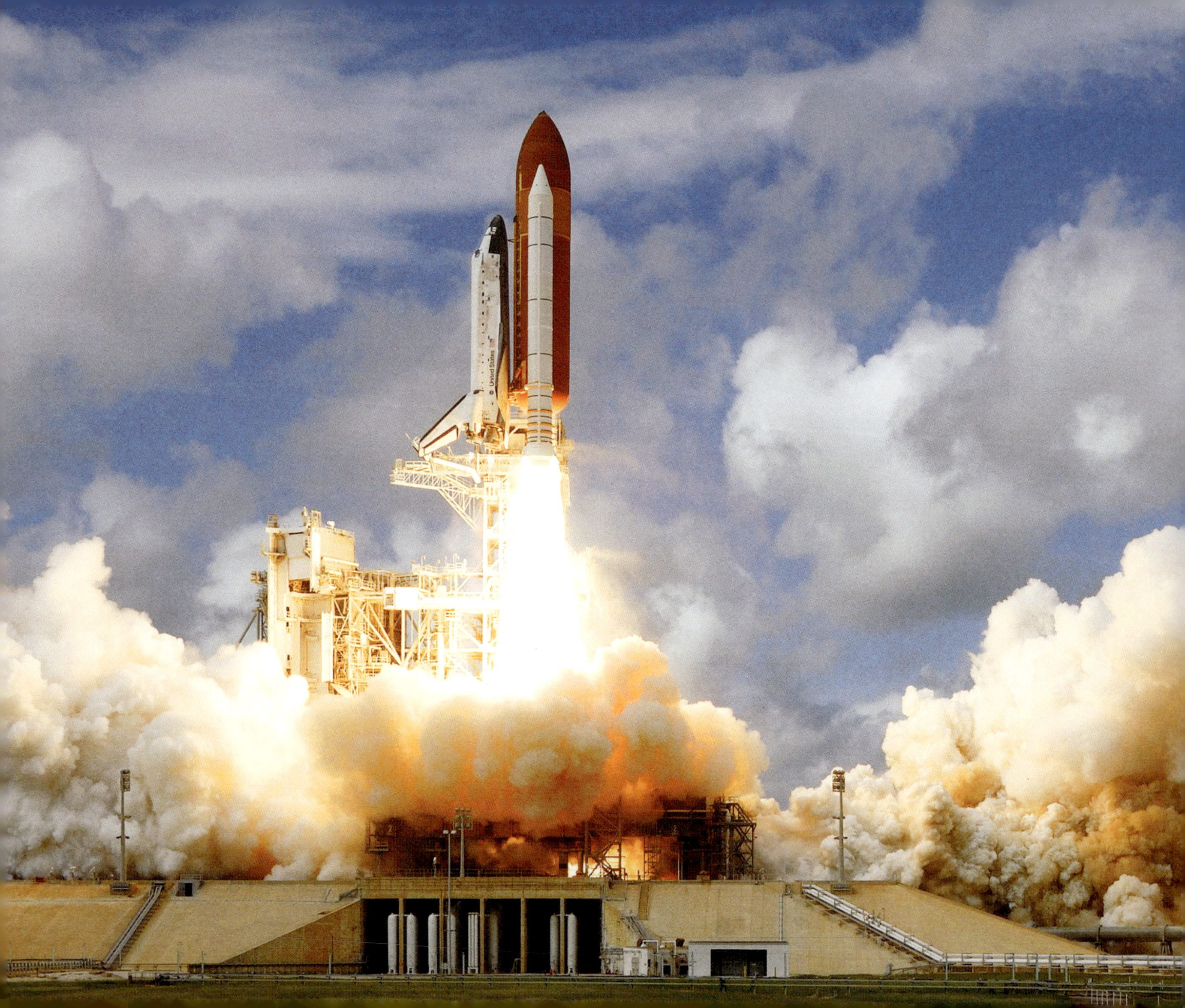

Energy

Liftoff!

Big blasts of energy from the rocket engines push the space shuttle up and lift it off of the ground.

The shuttle must fly extremely fast to go around the Earth. That's why the rocket engines keep blasting for liftoff. They stop once we reach space.

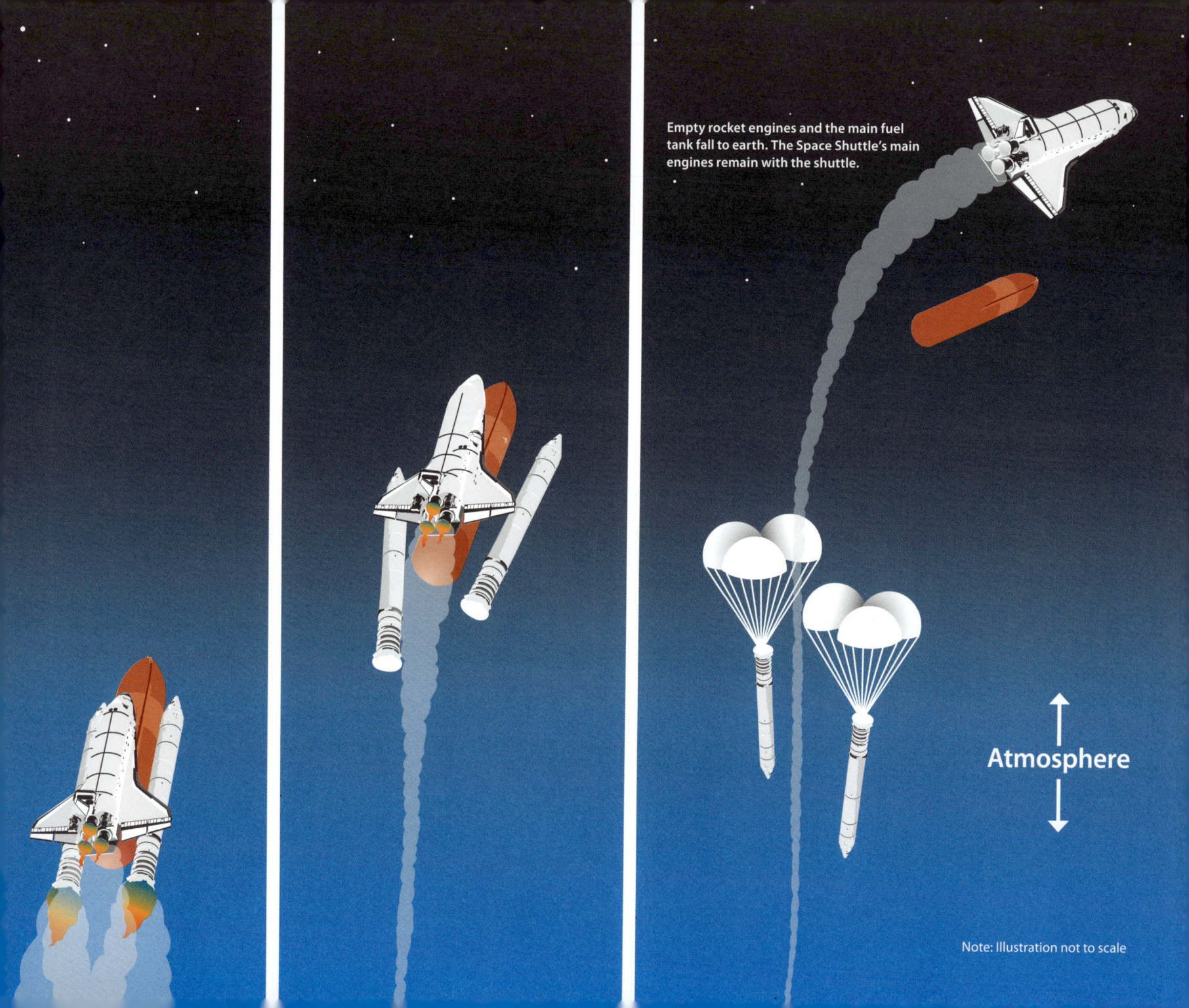

Empty rocket engines and the main fuel tank fall to earth. The Space Shuttle's main engines remain with the shuttle.

Atmosphere

Note: Illustration not to scale

Atmosphere

The **atmosphere** is the thin layer of air that wraps around the Earth. The space shuttle flies high above the atmosphere.

Can you guess how long it takes to get into space above the atmosphere? Only eight and a half minutes!

Along the way, the empty solid rocket engines and large fuel tank drop off and fall back to Earth.

shuttle orbit
atmosphere

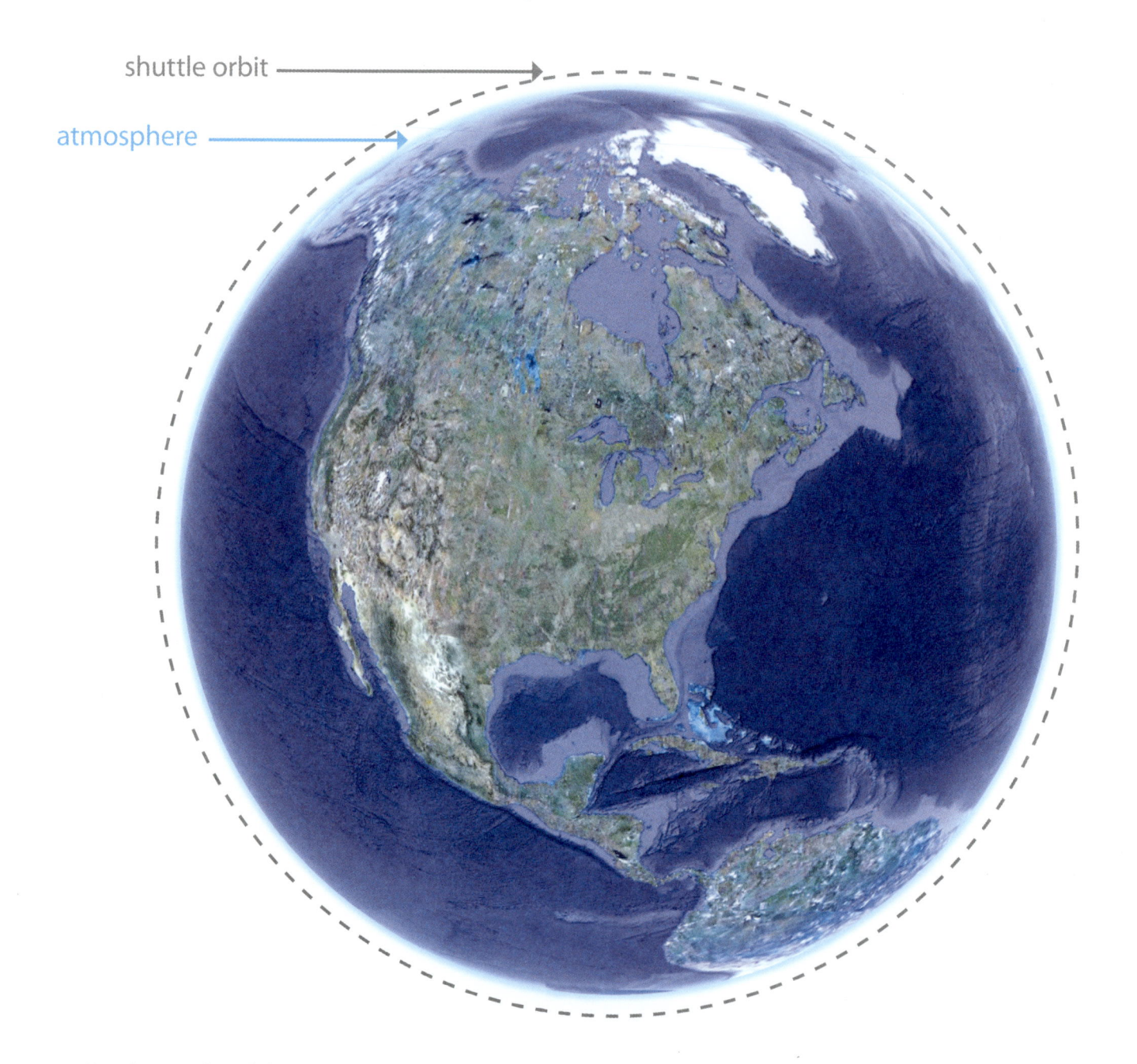

Note: the atmosphere "fades away" into space and does not have a definite edge. More than 99% of the atmosphere is within 25 miles of the earth's surface.

Orbit

It's amazing! The space shuttle can fly 250 miles above the ground, but compared to the size of the Earth, it doesn't seem very high at all.

Why doesn't the shuttle fall down? Unlike a ball that falls back down to the ground, the shuttle flies so fast it actually "falls" around the Earth.

The shuttle's path is called its **orbit**.

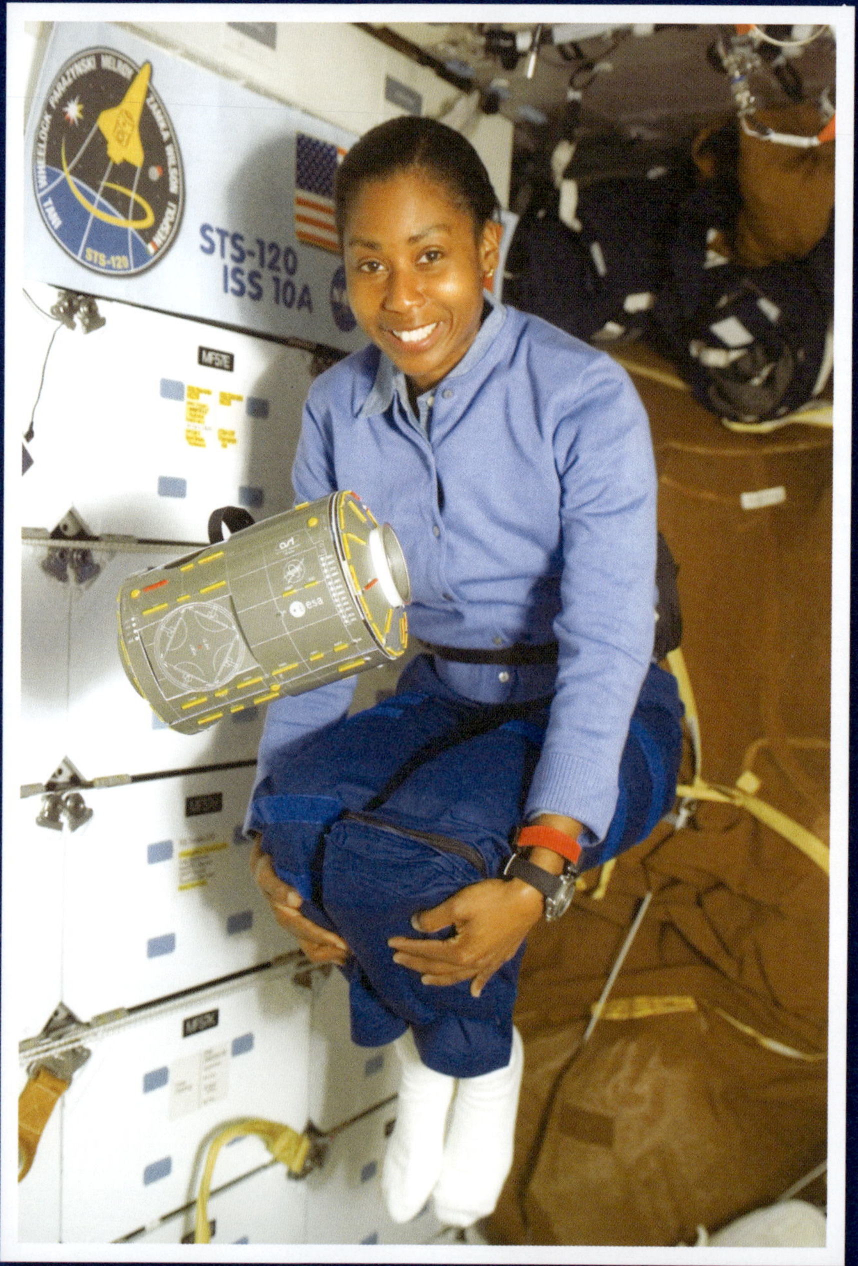
STS-120
ISS 10A
esa

Weightlessness

Once the space shuttle goes into orbit, everything floats. It doesn't matter if you are upside-down or right-side-up!

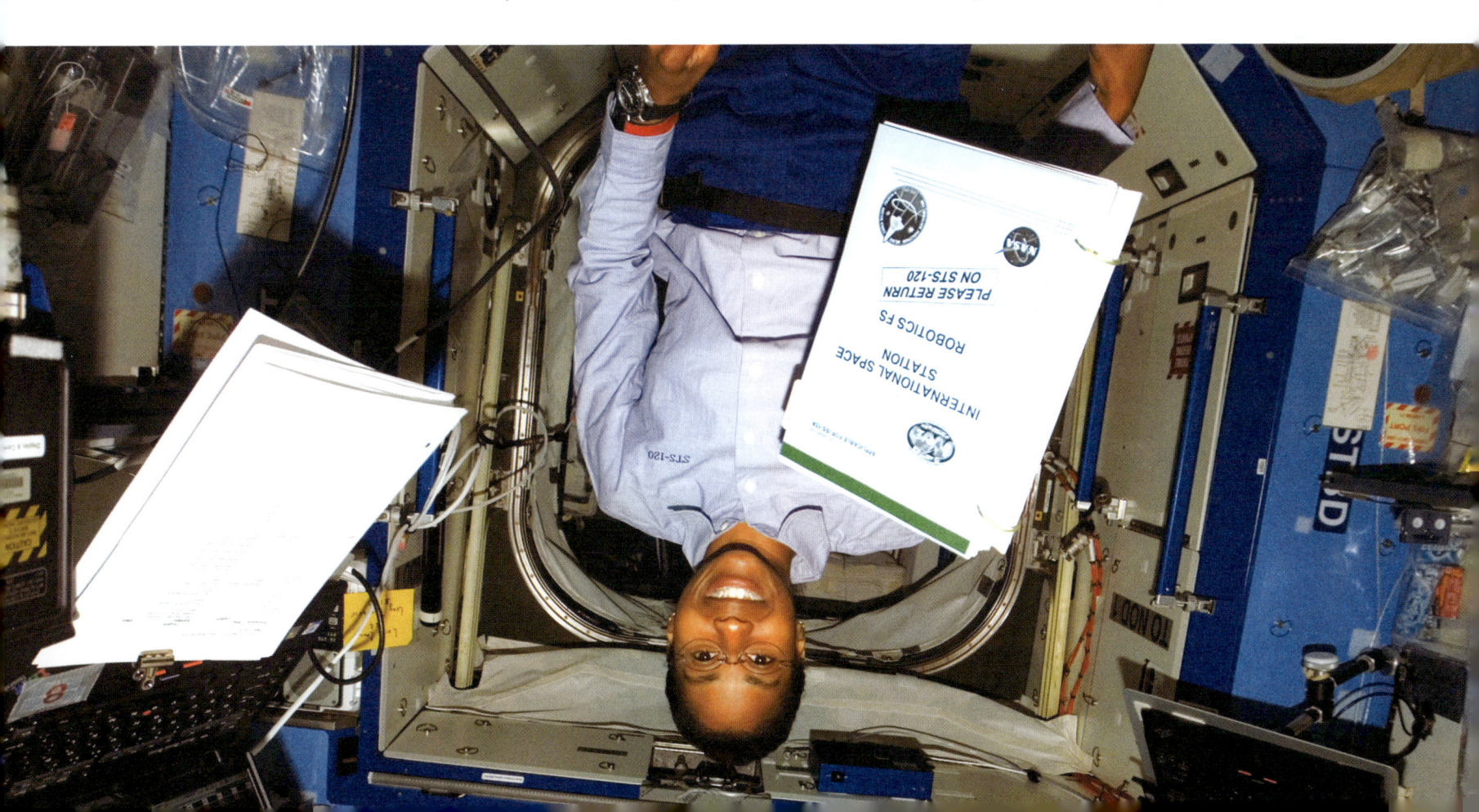

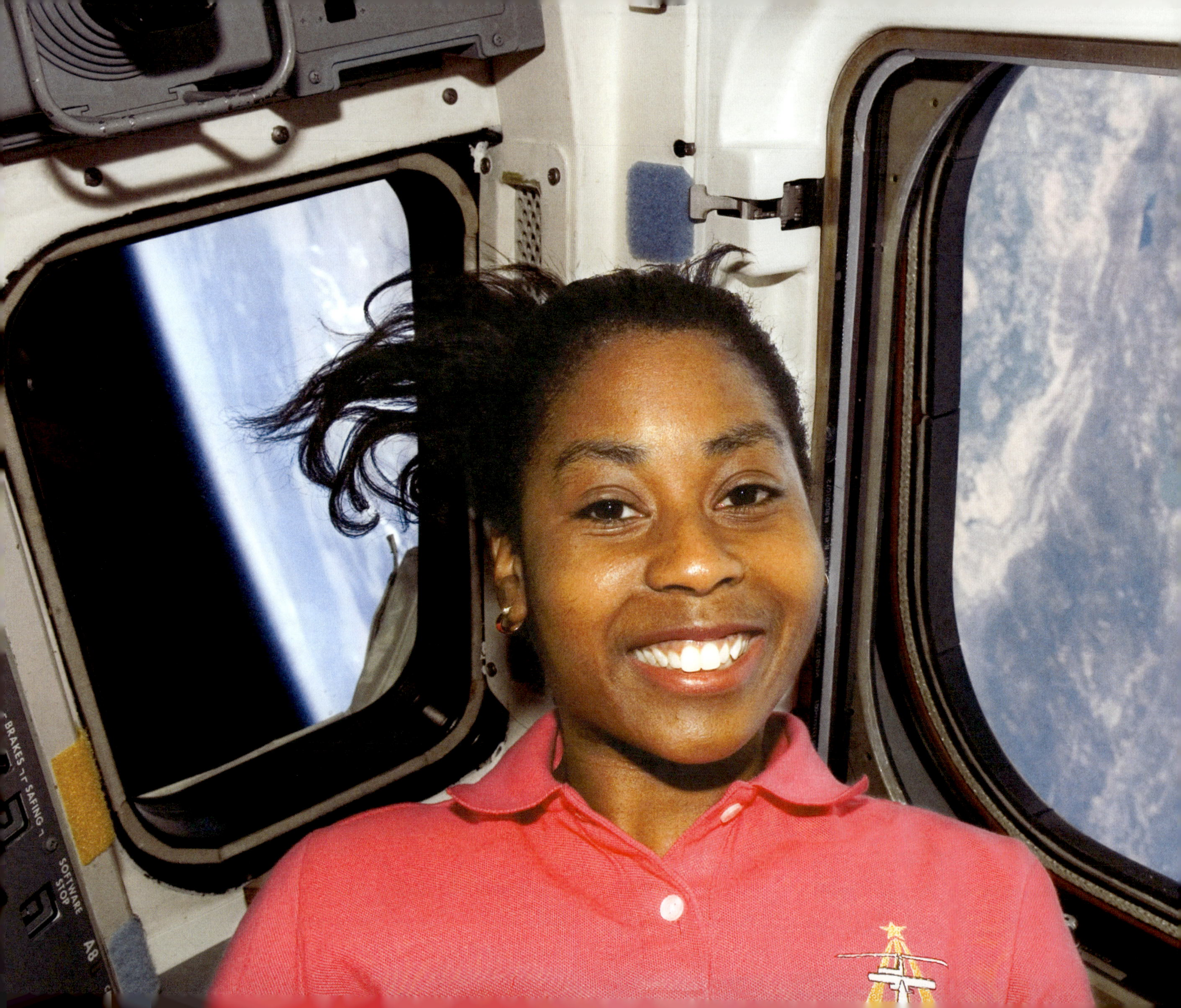
BRAKES
SAFING
SOFTWARE STOP
A8

Wonder

I can't believe this is a picture of me in space!

We all learn in school that the Earth is round, but it's amazing to actually see it with your own eyes!

Stephanie with Daniel Tani, flight engineer, checking instrumentation

Mission

Our **mission** included delivering a new section to the **International Space Station**.

Children from around the country submitted names for this new section. *Harmony* was the winning name.

My job was to use a special **robotic arm** to move *Harmony* from the cargo bay.

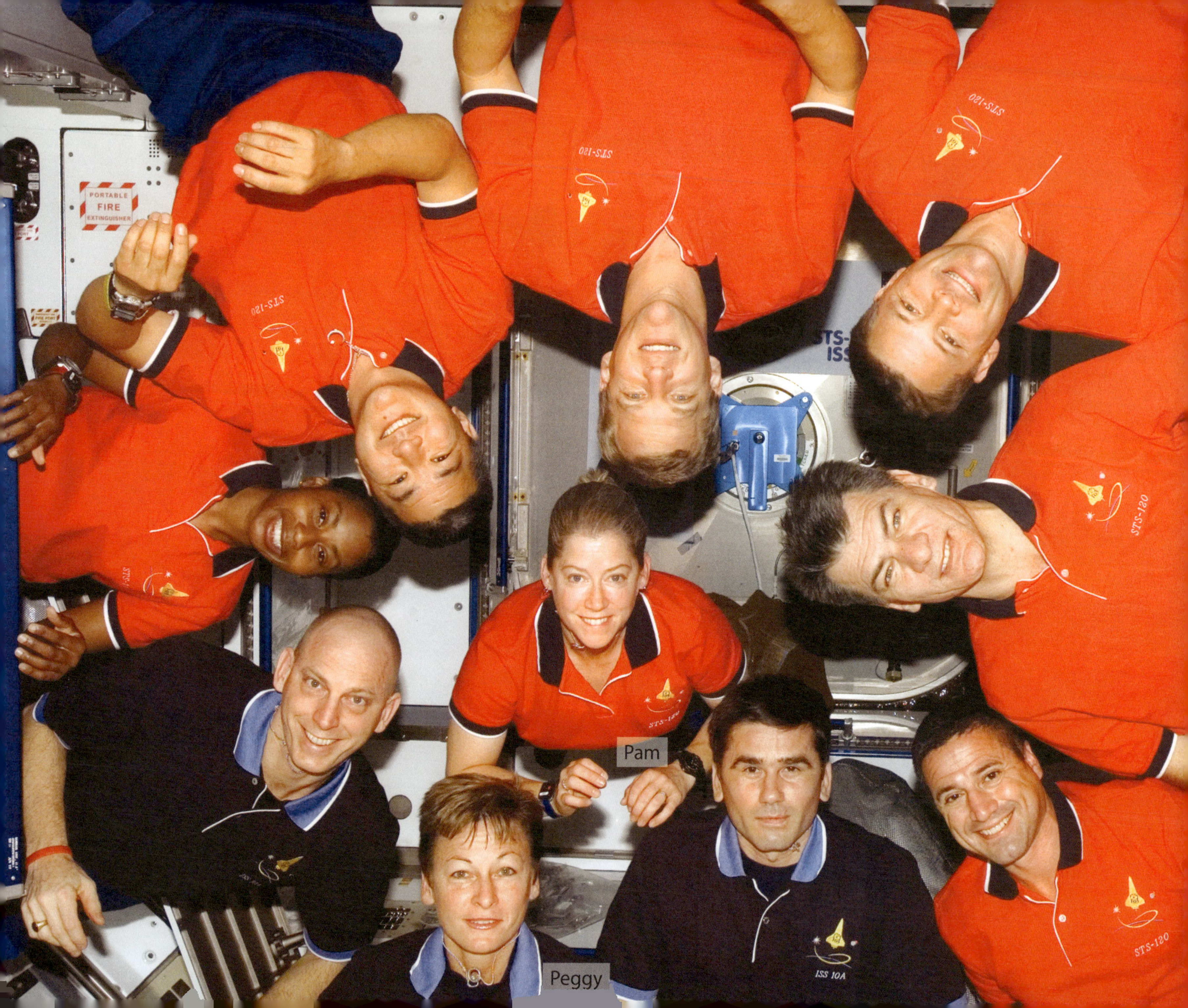
Pam
Peggy
PORTABLE FIRE EXTINGUISHER
STS-120
ISS 10A

Friendship

We made friends. Our crew of seven worked together with three astronauts already on the International Space Station. This group included astronauts from Italy and Russia.

Peggy was the commander of the International Space Station and Pam was the commander of the space shuttle.

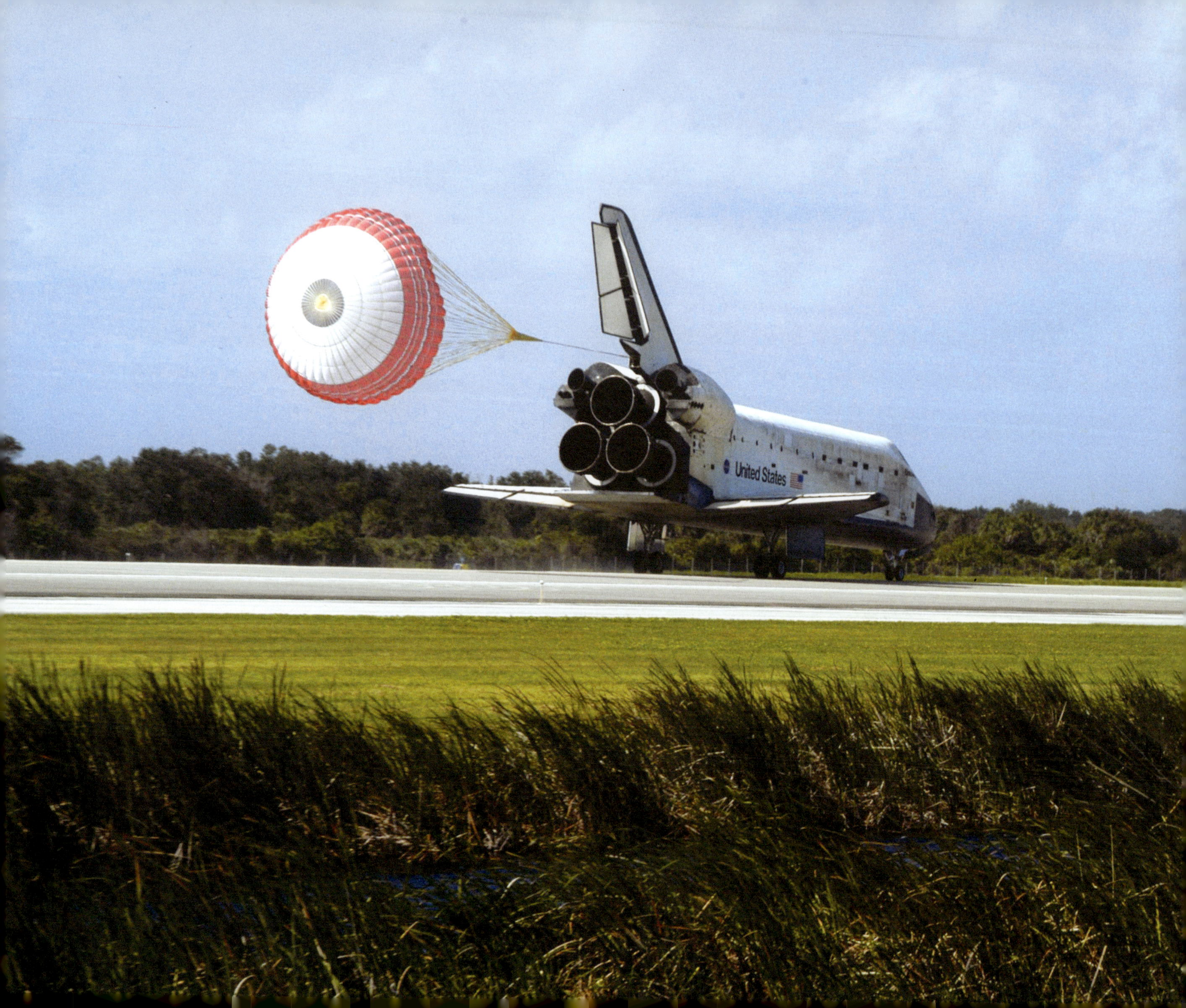
United States

Success

Returning to Earth is not easy.

The space shuttle moves much faster than a normal airplane. As it comes down, it rubs against the air and gets very, very hot.

The bottom of the shuttle is covered with special tiles to protect it from the heat. If the tiles weren't there, it would burn up!

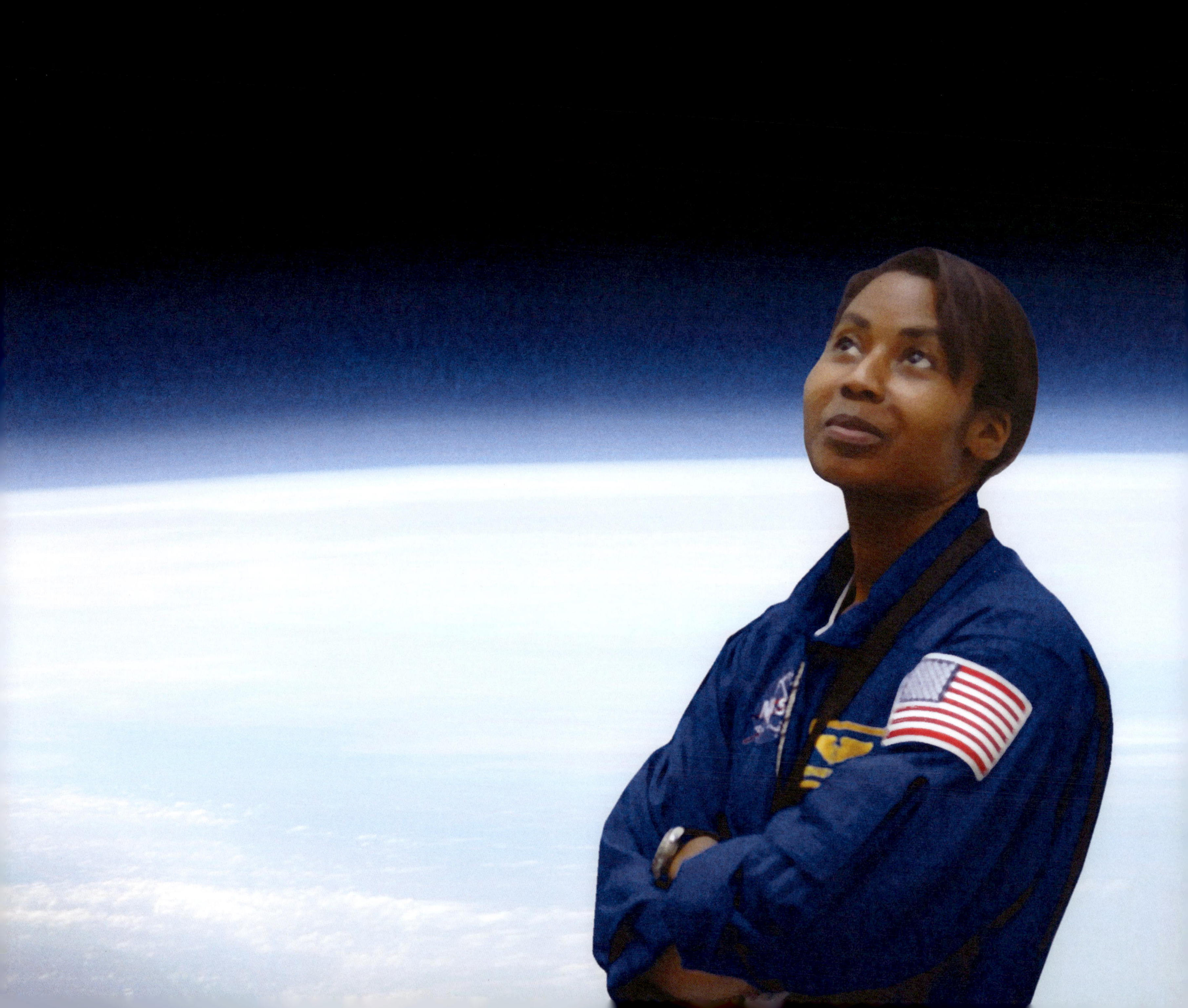

Dream!

I still dream.

I dream about the millions of stars and planets.

I dream about the different kinds of life forms that might be living out there somewhere.

What is your dream?

You should follow the dream that lies within your heart!

Vocabulary

astronaut
A person trained to travel in outer space.

astronomer
A person who studies objects in space such as planets and stars.

atmosphere
A thin layer of air around the Earth.

energy
The capacity to push an object over a distance.

enthusiasm
Having a lot of interest in something and being very excited about it.

International Space Station
A space station where humans live and work, built by a group of countries working together.

launch and entry suit
A special airtight suit that protects a person inside the space shuttle.

mission
A special job that is given to a person or group.

NASA
The "National Aeronautics and Space Administration," a United States agency that develops spacecrafts to explore outer space. Its goal is to learn about the Earth and space, and to develop human space flight.

orbit
The path an object follows as it moves in space. For example, the orbit of the space shuttle is around the Earth.

robotic arm
A large crane modeled after the human arm and controlled by an astronaut using 2 joysticks.

space shuttle
A spacecraft that carries cargo and people and has a section with wings that returns to the Earth like an airplane.

Index

A
astronaut   5, 9, 25, 30
astronomer   5, 30
atmosphere   15, 16, 30

E
energy   11, 13, 30
enthusiasm   5, 30

I
International Space Station   23, 25, 30

L
launch and entry suit   11, 30

M
mission   23, 31

N
NASA   9, 31

O
orbit   16, 17, 19, 31

R
robotic arm   23, 31

S
space shuttle   11, 13, 15, 17, 19, 25, 27, 31

Acknowledgments
Starfall would like to thank Stephanie D. Wilson for sharing her inspiring story with us.

About the Author
At age 9, young Stephen Schutz was still struggling to read. What came easily for some children required many more hours of Stephen's work, and he was consistently toward the bottom of his class in reading. Now a Ph.D. in physics and a successful publisher and artist, Dr. Schutz wants to make sure children in his situation today have resources that can help. He turned to the Internet and conceived Starfall.com, an online educational resource available to children the world over.

Photo Credits
Horsehead Nebula image on page 4, courtesy of NASA/ESA/NOAA. Photos on pages 2 and 6, courtesy of Stephanie D. Wilson. Photo of *Earth* on page 16, courtesy of GoogleEarth/TerraMetrics/DigitalGlobe. All other photos courtesy of NASA.